# YOUR KNOWLEDGE HAS VALUE

Clinton Kelly

# The utility of antiglobulin testing in blood group serology

GRIN Verlag

**Bibliografische Information der Deutschen Nationalbibliothek:**

Die Deutsche Bibliothek verzeichnet diese Publikation in der Deutschen National-
bibliografie; detaillierte bibliografische Daten sind im Internet über http://dnb.d-
nb.de/ abrufbar.

**Imprint:**

Copyright © 2013 GRIN Verlag GmbH
Druck und Bindung: Books on Demand GmbH, Norderstedt Germany
ISBN: 978-3-656-63449-2

**This book at GRIN:**

http://www.grin.com/en/e-book/271700/the-utility-of-antiglobulin-testing-in-blood-
group-serology

## GRIN - Your knowledge has value

Der GRIN Verlag publiziert seit 1998 wissenschaftliche Arbeiten von Studenten, Hochschullehrern und anderen Akademikern als eBook und gedrucktes Buch. Die Verlagswebsite www.grin.com ist die ideale Plattform zur Veröffentlichung von Hausarbeiten, Abschlussarbeiten, wissenschaftlichen Aufsätzen, Dissertationen und Fachbüchern.

**Visit us on the internet:**

http://www.grin.com/

http://www.facebook.com/grincom

http://www.twitter.com/grin_com

# JOMO KENYATTA UNIVERSITY

## OF

## AGRICULTURE AND TECHNOLOGY

**ASSIGNMENT: HML 3162 BLOOD TRANSFUSION PRACTICE**

# The Utility of Antiglobulin Testing In Blood Group Serology

**Mr. WIGINA R., Nyarambe**

**Date of submission:  28/10/2013**

# Table of contents

## Contents

Table of contents ........................................................................................................ 2

Introduction ............................................................................................................... 3

The Antiglobulin Test Systems ................................................................................. 3

    Liquid phase systems ............................................................................................. 3

    Column agglutination systems ............................................................................... 4

    Solid Phase ............................................................................................................ 4

The antiglobulin techniques and types ....................................................................... 4

    The direct antiglobulin test ................................................................................... 4

    The indirect antiglobulin test ................................................................................ 5

The Use of antiglobulin testing and reagents ............................................................ 5

    Blood Grouping .................................................................................................... 5

        Introduction .................................................................................................... 5

        Utility of Antiglobulin Testing ...................................................................... 5

    Detection of haemolytic disease of the Newborn/foetus (HDN/F) ......................... 5

        Introduction .................................................................................................... 5

        Utility of Antiglobulin test ............................................................................. 6

    Identification and screening of blood group system antigens/antibodies, Research and Education ........................................................................................................ 7

        Introduction .................................................................................................... 7

        Utility of antiglobulin testing ........................................................................ 8

    Detection and identification of transfusion reactions ............................................ 8

        Introduction .................................................................................................... 8

Utility of antiglobulin Testing ................................................................................................. 9

Other uses of antiglobulin testing other than blood group serology................................................. 9

REFERENCES ........................................................................................................................... 10

# The Utility of the Direct and Indirect Agglutination in blood group serology

## Introduction

The detection of reactions between antigen and antibody has been used to "phenotype" cells and to establish the presence of either antibody or antigen. Blood group antigens are either IgG or IgM. Though divalent, the IgG molecule is monomeric and the distance between two Fab regions is not generally enough to allow for direct agglutination. This therefore means that the detection of IgG reactions will have to be enhanced.

The most commonly employed techniques include the use of enzymes to cleave negatively charged particles on the surface of the red blood cells in order to reduce the negative charge and hence repulsion of the red cells. This then reduces the distance between cells and enables them to come together whence an agglutination reaction can be observed. Secondary antibodies may also be used to help in the detection of the reaction.

Apart from blood group serology, the detection of other human proteins which are capable of developing IgG antibodies and fixing complement can utilize this technique. Disease therapy monitoring in immunoglobulin therapies may also employ this technique.

## The Antiglobulin Test Systems

Test systems that have been used in the detection of serological reactions can be classified into three broad categories namely

### Liquid phase systems

This is the gold standard for detection of serologically significant reactions. The detection of reaction is by use of tubes or microtitre wells to visualize the reaction. There need be meticulous attention to the reactions and especially when the indirect antiglobulin test is performed and at the washing stage in particular.

## Column agglutination systems

This simple column test allow for the use of glass beads or a gel system in six columns. The gel or microbead system is formulated to allow the passage of unagglutinated cells to the bottom but not agglutinated cells. A positive reaction is thus characterised by agglutinates at the top of the column and a button of free red cells at the bottom. Reagent IgM or Antiglobulin can thus be added to type the reaction without need for washing.

## Solid Phase

In the solid phase system, a monolayer of cells on the plate surface signifies a positive reaction while a discrete button of cells is negative. The solid phase systems require accurate and standardized centrifugation steps. The solid phase techniques are thus very well adaptable to automated systems.

The use of Antiglobulin techniques first described by Carlo Moreschi in 1908 and rediscovered in 1945 by Robin Coombs, Rob Race, and Arthur Mourant, allowed the identification of many other blood group antigens in the decades that followed(Michael & Derwood 2009).

# The antiglobulin techniques and types

Today, three types of antiglobulin types are available in the market. These are either polyclonal or monoclonal. The use of polyclonal test sera is however not in common use. The antiglobulins available include;

a.  Anti-IgG antiglobulin this contains only anti IgG antibodies
b.  Anti-Complement C3d or C4d. These reagents contain only the monoclonal antibodies against complement which are used especially to detect complement activity without the undue influence of complement fixing antibodies
c.  Anti-IgG + anti-Complement. This is a blend which is used when the detection of complement fixing is desirable(Beck 2009).

There are two types of antiglobulin testing, the direct and the indirect antiglobulin tests.

## The direct antiglobulin test

The direct test is used to detect the antibodies that have reacted with the patient red cells in vivo. These maybe found coating red cells and may be associated with alloantibodies coating red cells in Haemolytic disease of the Newborn or autoantibodies in autoimmune haemolytic anaemia.

## The indirect antiglobulin test

The indirect test is used to detect the presence of antigens or antibodies when sensitized in vitro. This test is associated with antibody identification, typing and grouping.

The antiglobulin tests have been used widely in blood transfusion and transplantation. The following are typical uses of the tests.

# The Use of antiglobulin testing and reagents

## Blood Grouping

### Introduction

Many blood group system antigens have corresponding antibodies that are IgG. These are not able to directly agglutinate red cells. In effect therefore, although they coat the cells and indeed there are reactions, these are not visible. The addition of antiglobulins will therefore make visible the reactions that were otherwise not visible. Many antigens have been discovered by the use of these globulins and thus their characterization been completed.

Of the carbohydrate antigens, potent IgG anti Le[a] has been demonstrated.

### Utility of Antiglobulin Testing

The most commonly detected antigens/antibodies using the indirect antiglobulin tests are the protein antibodies. Naturally occurring IgG anti Duffy (Fy) –Kell (K &k and Js) –Kidd (Jk) and other protein antigens have been demonstrated using indirect antiglobulins(Klein & Anstee 2005).

When the nature of the antigen is such that it is embedded too deeply within the cell membrane, enzymes such as trypsin and papain may be used to digest the sialic acid layer thus reducing the negative charge between cells. This treatment makes it possible to reach these antigens. The antiglobulin is then able to bridge the distance between the IgG molecules and a visible agglutination is observed. An example of this is the P1 antigen. The determination of variant antigens such as the Rh D variants can also be done using the indirect antiglobulin test. It is essential to not that polyclonal antibodies are best used in such assays as opposed to monoclonal(Kulkarni et al. 2013)

## Detection of haemolytic disease of the Newborn/foetus (HDN/F)

### Introduction

This disease puzzled scientists in the late 19th and early 20th centuries. To date, the Rh haemolytic disease of the newborn remains the single most common and severe form of this disease. Intervention in the first pregnancy helps to reduce fatalities in the next to a great deal. It is therefore important to have tests that can be used to check the presence of either antibodies or antigens that may result in HDN/F.

**Maternal antibodies**(de Alarcon & Werner 2005)

| Associated with severe HDN | Usually associated with mild to moderate HDN | Not associated with HDN |
|---|---|---|
| RhD, Rhc, RhC, RhE, | ABO, | $*Fy^b$ |
| Kell 1 [K], Kell2 [k], $Js^b$, | $Kp^a$, $Kp^b$, LW | $*N$, $*s$, |
| $Fy^a$ | $Jk^a$, $Jk^b$, $Jk^3$, $Js^a$, | P1, |
| M, S, U | Ula (Kel 10], | $Le^a$, $Le^b$, |
| | $Fy^3$ | $Lu^a$, $Lu^b$ |

Adapted from neonatal hematology

*One reported case of mild or moderate HDN. $Js^a$ only mild cases of HDN/F, $Js^b$ moderate and severe cases reported.

Although the incidences of Rh HDN have decreased considerably over the years due to the development of the anti-RhD therapy, many other antibodies have been discovered to lead to HDN. The incidences of which have increased over the years.

The detection of these has greatly depended on the antiglobulin test among others. There are many other causes of neonatal jaundice and it would therefore be best to differentially diagnose what type of HDN so that the proper treatment is done.

Blood group HDN occurs due to fetomaternal hemorrhage or due to a prior transfusion of the mother with blood containing the same antigen as the fetus.

### Utility of Antiglobulin test

The detection of likelihood of HDN maybe first ascertained from maternal blood. Because the Mother must have been alloimmunized, it is possible to detect the culprit antibodies in maternal circulation.

This is done by the **indirect antiglobulin test**.

Maternal serum is screened for any antibodies that are likely to cause haemolytic disease to the fetus. This test although crucial, is limited to mothers who are known to be Rh Negative. This is especially so in resource poor settings. The practice thus precipitates a scenario where preventable HDN/F may not be detected due to none testing of the known potential blood group causes of HDN. This concern was addressed by a study group in turkey that assessed the risk of alloimmunization in pregnant women(Altuntas et al. 2013).

Maternal antibodies detectable by the indirect antiglobulin test include anti-Rh, anti-Kell, anti-Duffy, anti-MNSs and anti-Kidd(Hillyer et al. 2007).

The indirect antiglobulin test may also be performed using the amniotic fluid. Many or all of the antibodies detected here are IgG antibodies that are able to cross the placenta and cause a shortening of the fetal red cell lifespan.

Some other antibodies may also be detected in maternal serum during screening. If these antibodies are detected in the $37^0$ phase, then they should be viewed as potentially able to cause HDN/F otherwise it would be of no harm to ignore them. These include anti-P, anti-Lutheran, anti-Lewis and anti-Sid(Hillyer et al. 2007).

Other antibodies that are detectable in maternal antibody screens are harmless antibodies of the JMH, anti-Chido-Rodgers and anti-Knops antibodies(Daniels 2002)

Indirect antiglobulin tests may also be done on antibodies obtained from the eluate from the amniotic fluid after an amniocentesis.

**The direct antiglobulin test** is used in the detection of fetal/neonatal red cells coated with alloantibodies against it. These are again IgG antibodies.

The direct detection of these is a significant clue to the presence of active haemolytic disease.

Fetal cells may be obtained from the amniotic fluid during amniocentesis. The cells are then washed and a direct test performed to detect whether the cells may have been coated by the maternal antibodies.

The detection of antibody-coated fetal red cells is an important step in the evaluation of fetal survival in known fetomaternal incompatibilities. The decision to either terminate and/or prepare for exchange transfusion rests both on the maternal antibody titre and the degree of coating of the cells. In addition, the decision for intrauterine transfusion may lead to survival of otherwise infants who would not survive at the ages when these are detected(Hillyer et al. 2002).


## Identification and screening of blood group system antigens/antibodies, Research and Education

### Introduction

The red blood cell has been an object of study for a long time now. The study of the RBC surface has been both interesting and dynamic. Since Landsteiner discovered the blood groups in early $20^{th}$ century, many more blood group systems have been found.

It is impressive the number of antigens that have been discovered so far. These have been very instrumental in detailing the structure and function of the red blood cell and its interaction with its microenvironment.

The identification and screening of blood group systems may be done by use of the direct blood grouping techniques. However, there are many of the antigens that are either deeply embedded into the cell membrane structure or others whose antibodies are weak IgG.

**Utility of antiglobulin testing**

To be able to screen for these, various techniques are usually employed by the transfusion scientist. These include the antiglobulin testing, elution and enzyme treatment among others.

A classic example of the use of indirect antiglobulin testing for the detection/screening of blood group systems is the test for weak D also previously referred to as the $D^U$ test. In this test all Rh-D negative blood by direct cell grouping are incubated with the anti-D antibody. The anti-globulin is then added to determine whether there is a weak antigen to which the antibodies may have bound(Daniels & Bramilow 2007).

Many blood group system antigens have been demonstrated by the use of the indirect antiglobulin test. These include the MER2 antigens of the RAPH blood group system which was the first red cell surface polymorphism to be demonstrated by monoclonal antibody testing. Andresen documented that the $Le^a$ and $Le^b$ cannot usually be detected on cord blood by direct agglutination and indirect antiglobulin testing or agglutination of enzyme treated cells is employed(Andresen 1948).

The antiglobulin testing is therefore also used to demonstrate the presence of these blood group system substances in a class setting. Where the presence of the antigen or antibody is to be shown, in research, the antiglobulin testing is designed to help discover antigens and their antibodies.

The identification of antibody type is also done by the use of class specific antiglobulin reagents as an indirect antiglobulin test(Klein & Anstee 2005).

**Detection and identification of transfusion reactions**

### Introduction

Transfusion related reactions (adverse effects of transfusion) have been recognized since the first form of reactions was recognized by Dr. Jean Baptiste Dennis in the 17th century and conclusively reported by Levine in 1939(Hillyer et al. 2007).

Transfusion related complications arise from the transfusion of blood which has antigens or antibodies corresponding to antibodies or antigens present in the recipient. The transfer of antibodies will normally result in mild transfusion reactions although severe reactions and anaphylatoxic reactions have also been reported. The transfer of antigens on the other hand will result in mild to severe transfusion reactions. This happens either by complete and immediate destruction of the transfused cells or a partial/delayed and more gradual destruction.

**Utility of antiglobulin Testing**

Potential transfusion reactions may be first alleviated by the use of antiglobulin testing. This is achieved during the anti-globulin phase of compatibility testing(Taylor et al. 2011). The crossmatch technique employs the indirect antiglobulin test to detect whether there are antibodies in the patient serum that are able to react with donor cells and if there are donor antibodies (non-leucoreduced fresh whole blood transfusions) that are able to react with patient cells to cause adverse reactions.

The limitation of crossmatch is that it can only detect antibodies present in the serum only at the time of testing. This therefore means that antibodies developed post-transfusion are undetectable and are therefore able to cause transfusion reaction.

When a transfusion reaction occurs, it is possible detect whether there are antibodies in the blood causing transfusion reaction. Patient serum bay be retested and an indirect antiglobulin test done to ascertain if the patient is producing antibodies. The patient red cells may also be assayed in a direct antiglobulin test system to detect whether the transfused or the patient red cells are coated by allospecific antigens.

It is important to be able to distinguish between autoantibodies and allospecific antibodies in order to ascertain the occurrence of the blood transfusion reaction.

## Other uses of antiglobulin testing other than blood group serology

It is prudent also to mention that antiglobulin tests have been employed in tests for other conditions in haematology.

These include

a. Drug induced haemolytic disease e.g. in the Penicillins and Cephalosporins
b. Acquired immune haemolytic disease
c. Other human proteins that are implicated in disease e.g. fibrin degradation products in coagulation studies

# REFERENCES

De Alarcon, P.A. & Werner, E.J. eds., 2005. *Neonatal Hematology* 1st ed., Cambridge: Cambridge University Press.

Altuntas, N. et al., 2013. The risk assessment study for hemolytic disease of the fetus and newborn in a University Hospital in Turkey. *Transfusion and apheresis science : official journal of the World Apheresis Association : official journal of the European Society for Haemapheresis*, 48(3), pp.377–80. Available at: http://www.ncbi.nlm.nih.gov/pubmed/23619329 [Accessed August 29, 2013].

Andresen, P.H., 1948. The blood group system L: a new group L2 . A case of epistasy within the blood groups. *Acta Path Microbiol scandinavia*, 25, pp.728–31.

Beck, N., 2009. *Diagnostic Hematology* 1st ed., London: Springer.

Daniels, G., 2002. *Human Blood Groups* 2nd ed., oxford: Blackwell publishing.

Daniels, G. & Bramilow, I., 2007. *Essential Guide to Blood groups* 1st ed., Massachusetts: Blackwell publishing.

Hillyer, C.D. et al. eds., 2007. *Blood Banking and Transfusion Medicine (Basic Prinnnciples & Practice)* 2nd ed., Philadelphia: Elsevier Academic Press.

Hillyer, C.D., Strauss, R.G. & Luban, N.L.C. eds., 2002. *Pediatric Transfusion Medicine* 1st ed., New York: Elsevier Academic Press.

Klein, H.G. & Anstee, D.J., 2005. *Mollison's blood transfusion in clinical medicine* 11th ed., Massachusetts: Blackwel Science.

Kulkarni, S., Kasiviswanathan, V. & Ghosh, K., 2013. A simple diagnostic strategy for RhD typing in discrepant cases in the Indian population. *Blood transfusion = Trasfusione del sangue*, 11(1), pp.37–42. Available at: http://www.pubmedcentral.nih.gov/articlerender.fcgi?artid=3557475&tool=pmcentrez&rendertype=abstract [Accessed August 29, 2013].

Michael, M.F. & Derwood, P.H. eds., 2009. *Practical Transfusion Medicine* 3rd ed., west sussex, UK: John Wiley & Sons.

Taylor, J. et al., 2011. Multi-centre evaluation of pre-transfusional routine tests using 8-column format gel cards (DG Gel®). *Transfusion medicine (Oxford, England)*, 21(2), pp.90–8. Available at: http://www.ncbi.nlm.nih.gov/pubmed/21118316 [Accessed August 16, 2012].